5th Grade Math
Volume 5

© 2013 OnBoard Academics, Inc
Newburyport, MA 01950
800-596-3175
www.onboardacademics.com

ISBN: 978-1494857240

Table of Contents

One Step Equations

Key Vocabulary

equation

variable

balancing equations

What is the value of each figure?

$5 + \bigstar = 12$ $\blacklozenge + 3 = 15$

$\blacktriangle - 4 = 5$ $\bullet - 13 = 3$

$\bigstar =$ [ ]

$\blacklozenge =$ [ ]

$\blacktriangle =$ [ ]

$\bullet =$ [ ]

Complete the equations.

$5 + \bigstar = 12$ $\bigstar = \boxed{7}$

$\diamondsuit + 3 = 15$ $\diamondsuit = \boxed{12}$

$\triangle - 4 = 5$ $\triangle = \boxed{9}$

$\bigcirc - 13 = 3$ $\bigcirc = \boxed{16}$

$\diamondsuit + \boxed{} - \triangle = \boxed{15}$

$\bigcirc - \triangle = \boxed{}$

Expressions and Equations

Study the illustrations below to learn the difference between expressions and equations.

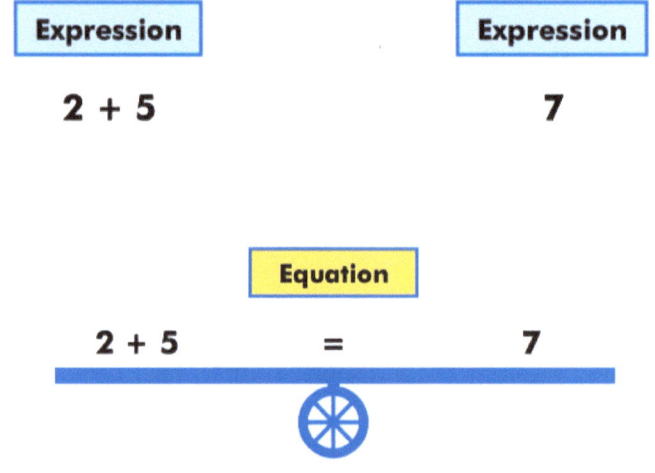

Expression	Expression
2 + 5	7

Equation

2 + 5   =   7

Sort the expressions and the equations.

Expression	Equation

$8 - 4 = 4$	$x - 9 = 4$	$4x$
$2 + 3 = 5$	$7 + 2$	$x + 3 = 9$

_____

Write and equation for these expressions.

Javier has eight dollars.	☐
He spends three dollars on a latté.	☐
How many dollars does he have left?	☐

Solving Equations

$$x + 3 = 8$$

$$x + 3 - 3 = 8 - 3$$

$$x = 5$$

The scale is modeling X + 3 = 8.

To solve the equation remove three from each side of the equal sign. Use the model and cross off boxes to model the answer.

Practice solving equations.

Hint

$$x + 3 = 8$$

$$x + 3 - 3 = 8 - 3$$

$$x = 5$$

CHECK $5 + 3 = 8$

_____

$$x + 1 = 7$$

$$=$$

$$x =$$

CHECK $=$

$$y - 6 = 8$$

$$=$$

$$y =$$

CHECK $=$

Solving Multiplication and Division Equations

Study the problem below to discover how to solve multiplication and division equations.

Matteus, James and Alison each have x marbles. When they put them together in a pile, there are 15 marbles.

How many marbles does each person have?

$$3x = 15$$

$$3x \div 3 = 15 \div 3$$

$$x = 5$$

$$3(5) = 15$$

$$3x = 15$$

$$\frac{3x}{3} = \frac{15}{3}$$

$$x = 5$$

$$3(5) = 15$$

Check

Practice solving multiplication and division problems.

$$4x = 40$$

$$\frac{n}{3} = 7$$

=

=

$x =$

$n =$

CHECK | =

CHECK | =

Name _____

One Step Equations Quiz

(1) True or false? An equation must have an = sign.

(2) x = ?

    A 4

    B 5

    C 6

    D 7

(3) Solve the equation $x - 9 = 14$

(4) Solve the equation $\frac{n}{4} = 8$

The Coordinate Plane

Key Vocabulary

coordinate plane

quadrant

origin

ordered pair

Find the ordered pairs.

Hint

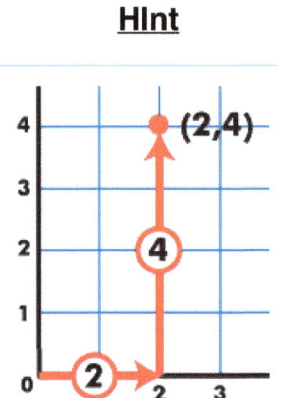

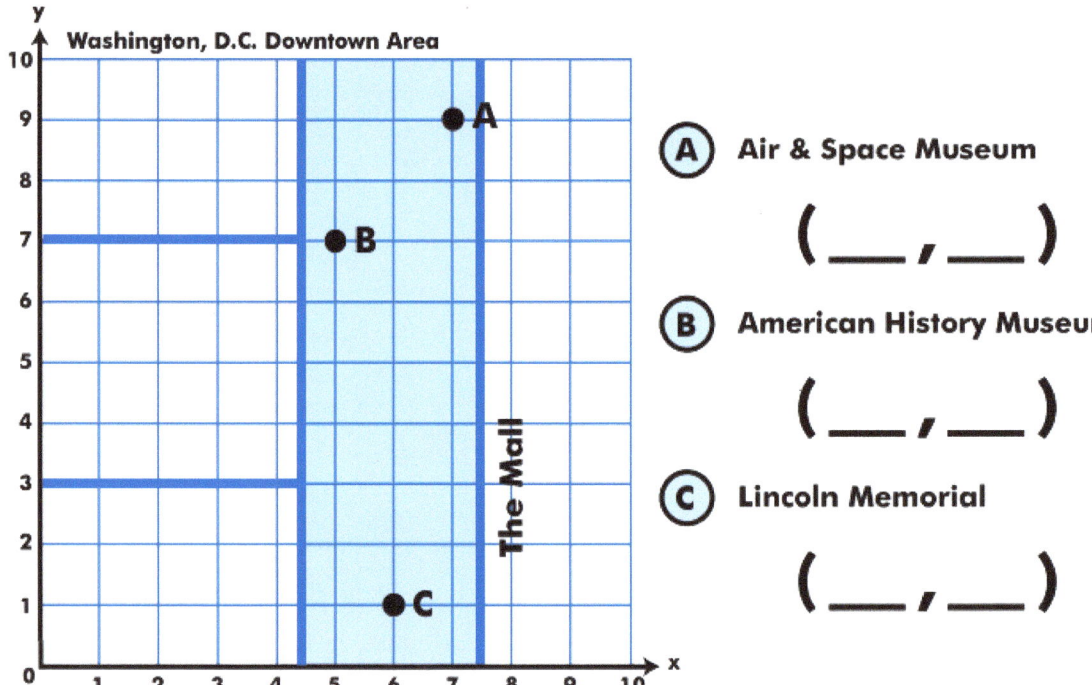

(A) **Air & Space Museum**

$$(\underline{\quad} , \underline{\quad})$$

(B) **American History Museum**

$$(\underline{\quad} , \underline{\quad})$$

(C) **Lincoln Memorial**

$$(\underline{\quad} , \underline{\quad})$$

Plot the ordered pairs.

Can you guess which monuments are (6,6) and (2,5)?

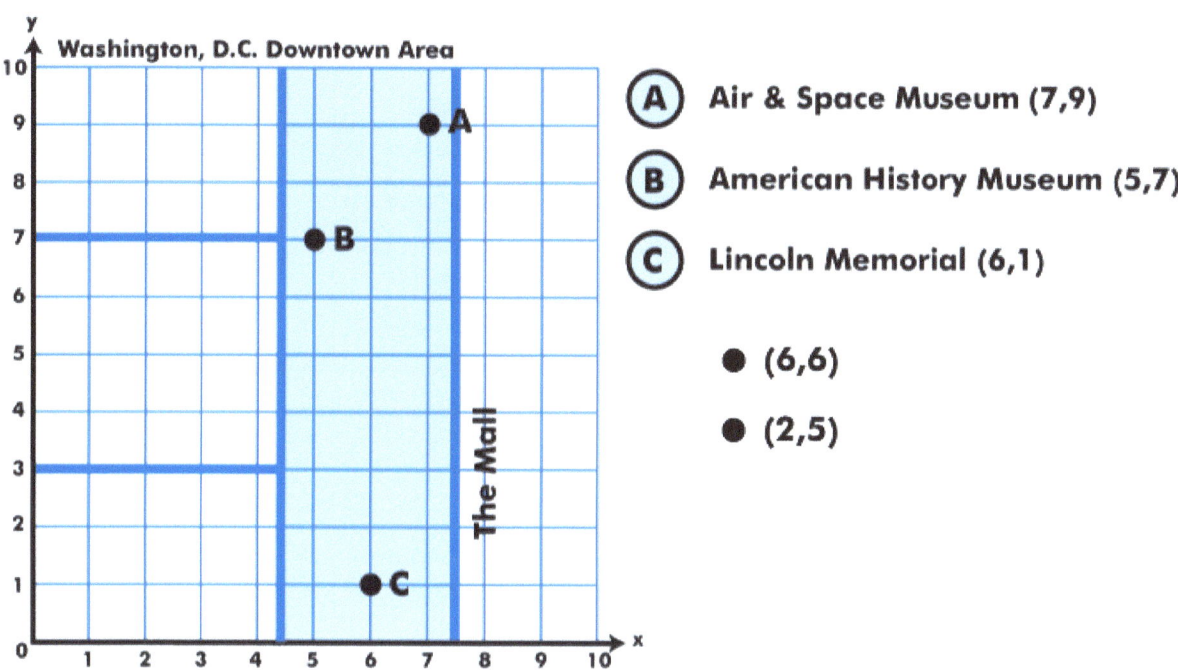

A   Air & Space Museum (7,9)

B   American History Museum (5,7)

C   Lincoln Memorial (6,1)

● (6,6)

● (2,5)

(6,6) is the Washington Monument

(2,5) is the White House

Find the ordered pairs.

Hint

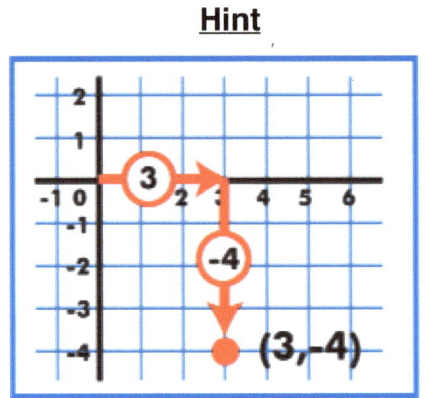

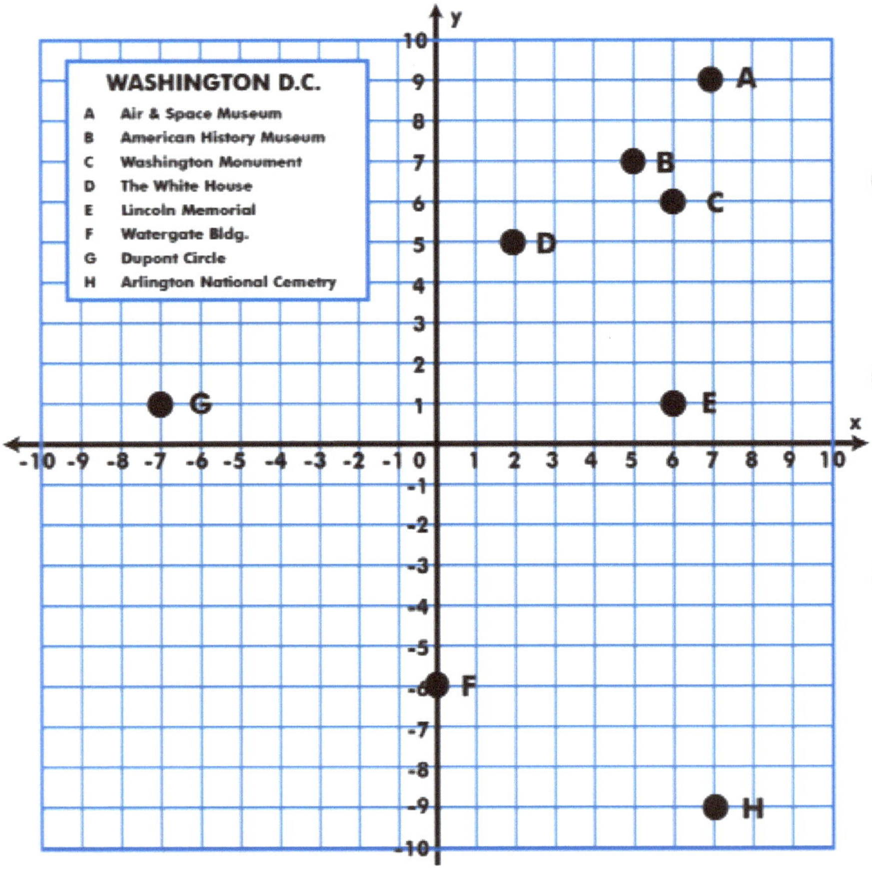

3
-4
(3,-4)

WASHINGTON D.C.
A Air & Space Museum
B American History Museum
C Washington Monument
D The White House
E Lincoln Memorial
F Watergate Bldg.
G Dupont Circle
H Arlington National Cemetry
A
B
C
D
E
F
G
H
x
y

(F) Watergate Building

(___ , ___)

(G) Dupont Circle

(___ , ___)

(H) Arlington Cemetry

(___ , ___)

Circle the image with the quadrants labeled correctly?

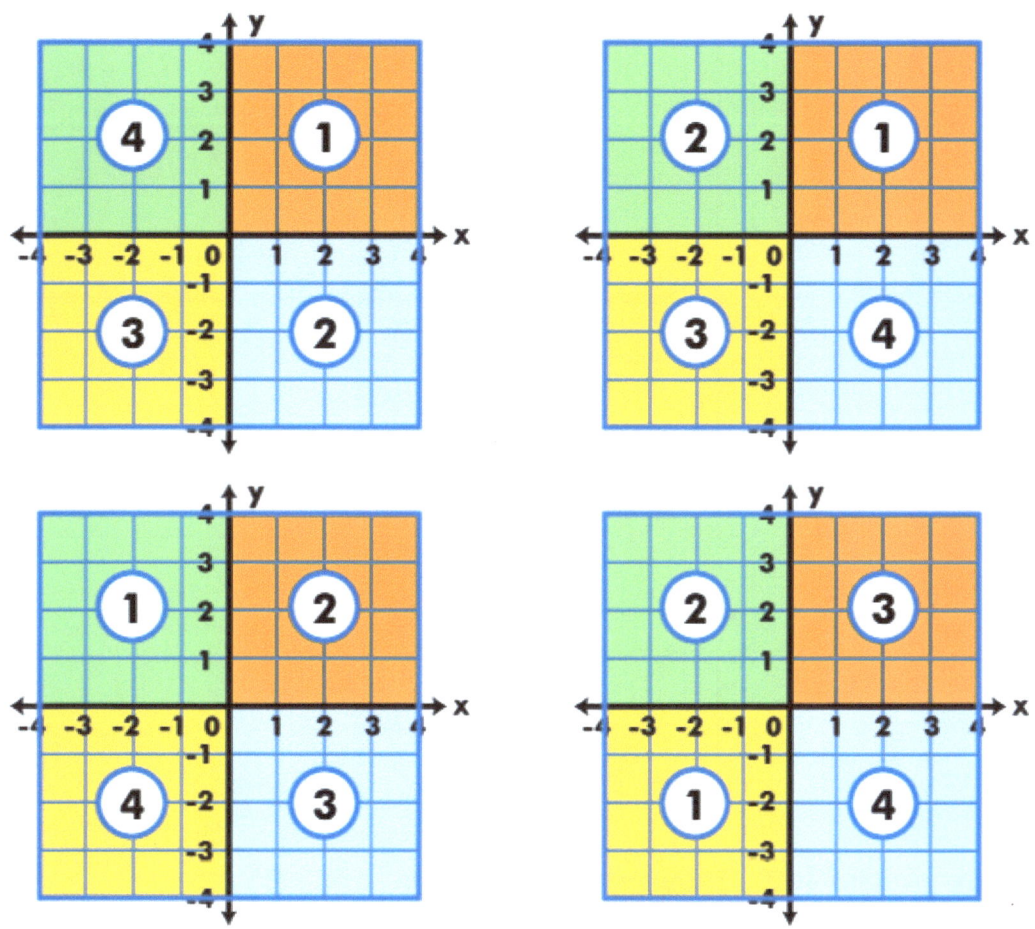

Write the ordered pair for A, B, and C.

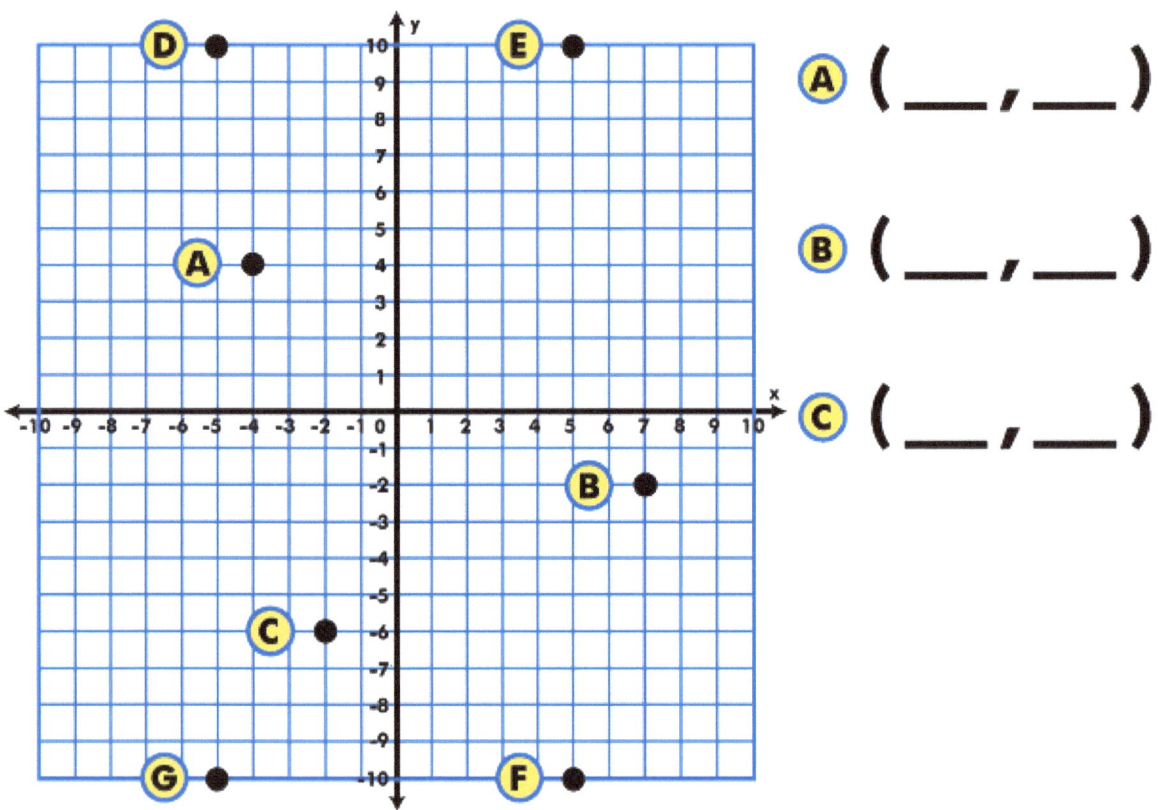

A (__ , __)

B (__ , __)

C (__ , __)

Which point has the coordinates (-5,10)?

Plot the points for A, B and C and then find an ordered pair for D to form a parallelogram.

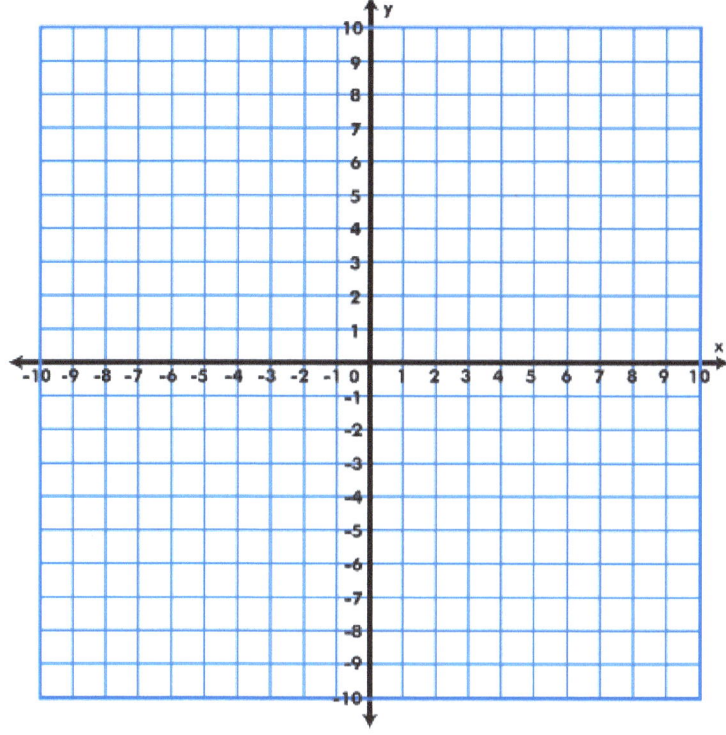

A (-5,-4)   ●

B (5,-4)   ●

C (0,2)   ●

D (__,__)   ●

Graphing Functions on the Coordinate Plane
Draw the points on the coordinate plane.

$y = 2x + 2$

x	y
1	4
2	6
3	8
4	10

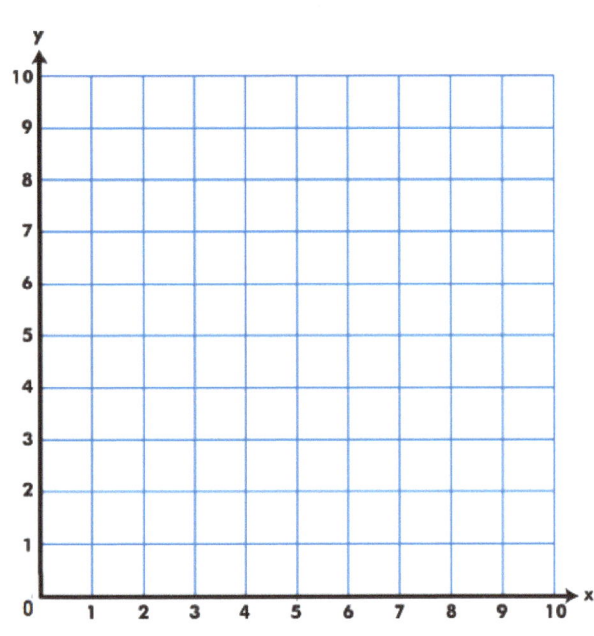

_____

Complete the graph for $y = 3x + 1$

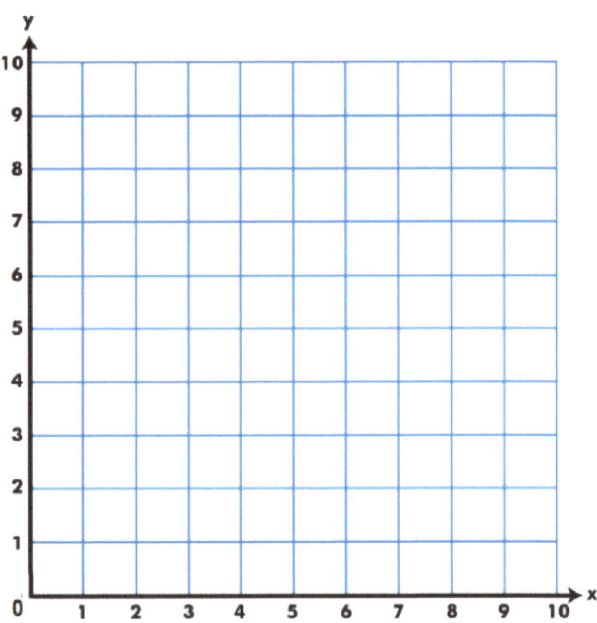

Name: _____

The Coordinate Plane Quiz

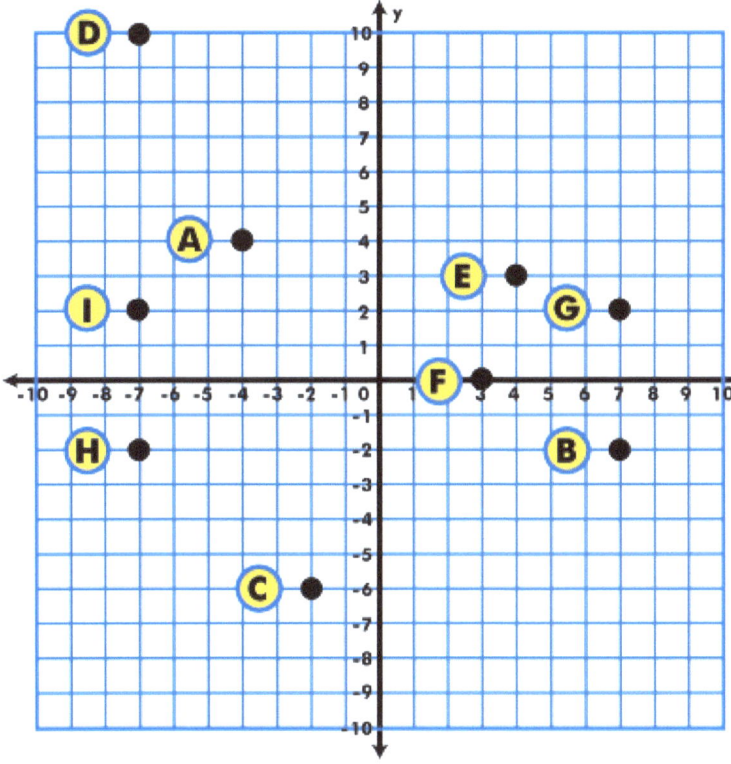

1. Which point has the coordinates (7,-2)?

2. Which point has the coordinates (-7,2)?

3. What is the y-coordinate of point F?

4. What is the y-coordinate of point E?

5. What is the x-coordinate of point E?

Properties of Quadrilaterals

Key Vocabulary

quadrilateral

parallelogram

trapezoid

rhombus

congruent

adjacent

Quadrilaterals

Polygon: a plane figure with at least three straight sides and angles.

Quadrilateral: a closed four-sided figure having four straight sides

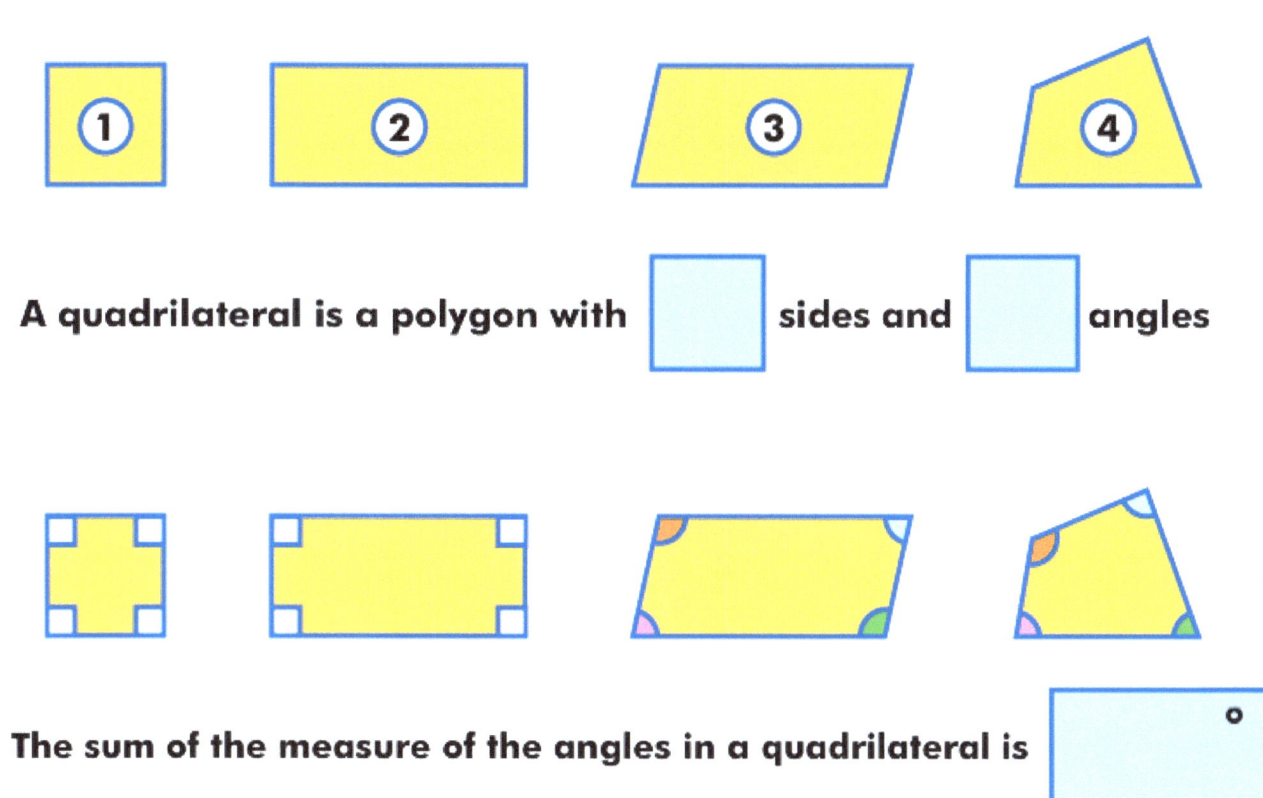

A quadrilateral is a polygon with [ ] sides and [ ] angles

The sum of the measure of the angles in a quadrilateral is [°]

Classifying Quadrilaterals: Parallelograms.

Step 1

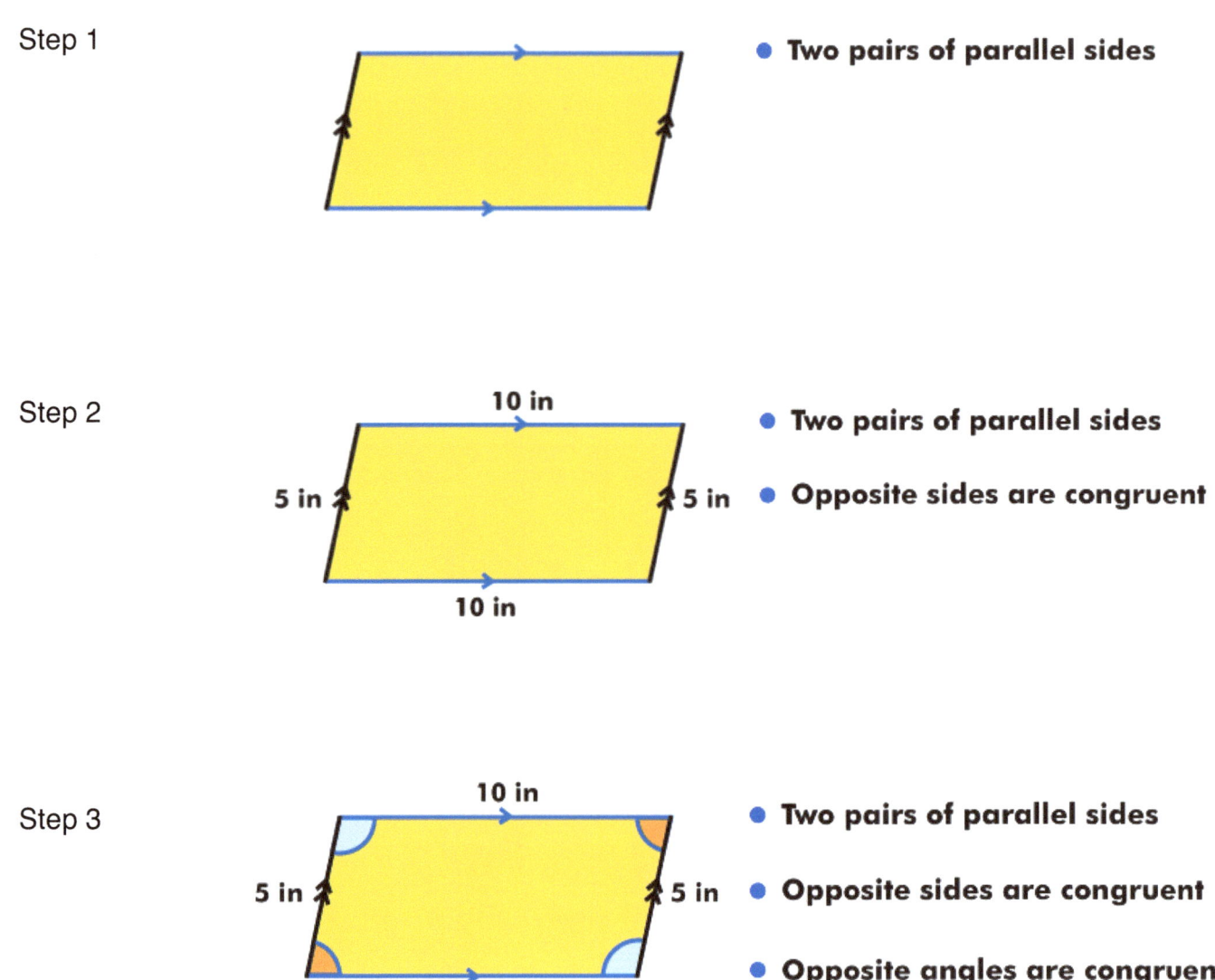

• **Two pairs of parallel sides**

Step 2

• **Two pairs of parallel sides**

• **Opposite sides are congruent**

Step 3

• **Two pairs of parallel sides**

• **Opposite sides are congruent**

• **Opposite angles are congruent**

Congruent: Same size and shape

Label the quadrilaterals and them sort them according to if they area also a parallelogram.

Use the three step classifying steps from the previous page.

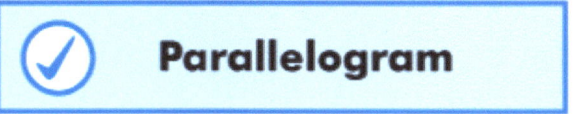

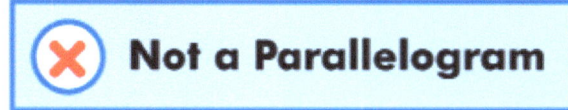

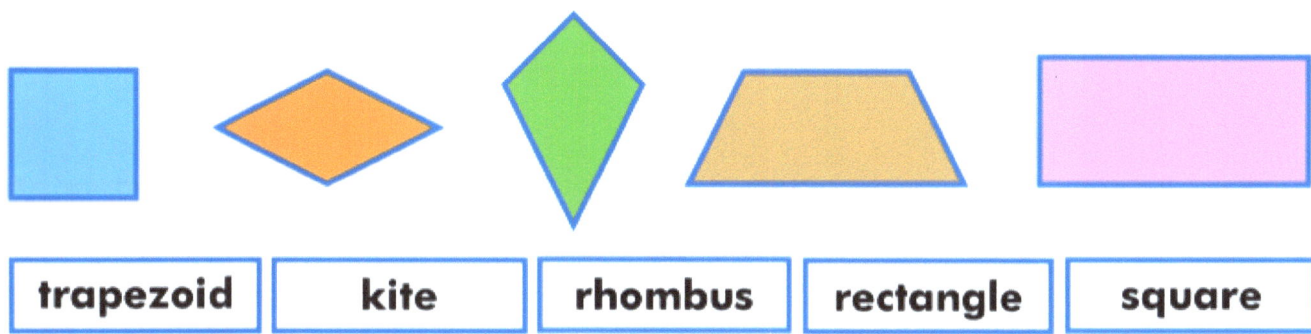

Square: a plane figure with four equal straight sides and four right angles

Rectangle: a plane figure with four straight sides and four right angles

Parallelogram: a plane figure with opposite sides being parallel

Rhombus: a parallelogram with opposite equal acute angles, opposite equal obtuse angles, and four equal sides.

Trapezoid: a quadrilateral with only one pair of parallel sides

Classify these quadrilaterals.

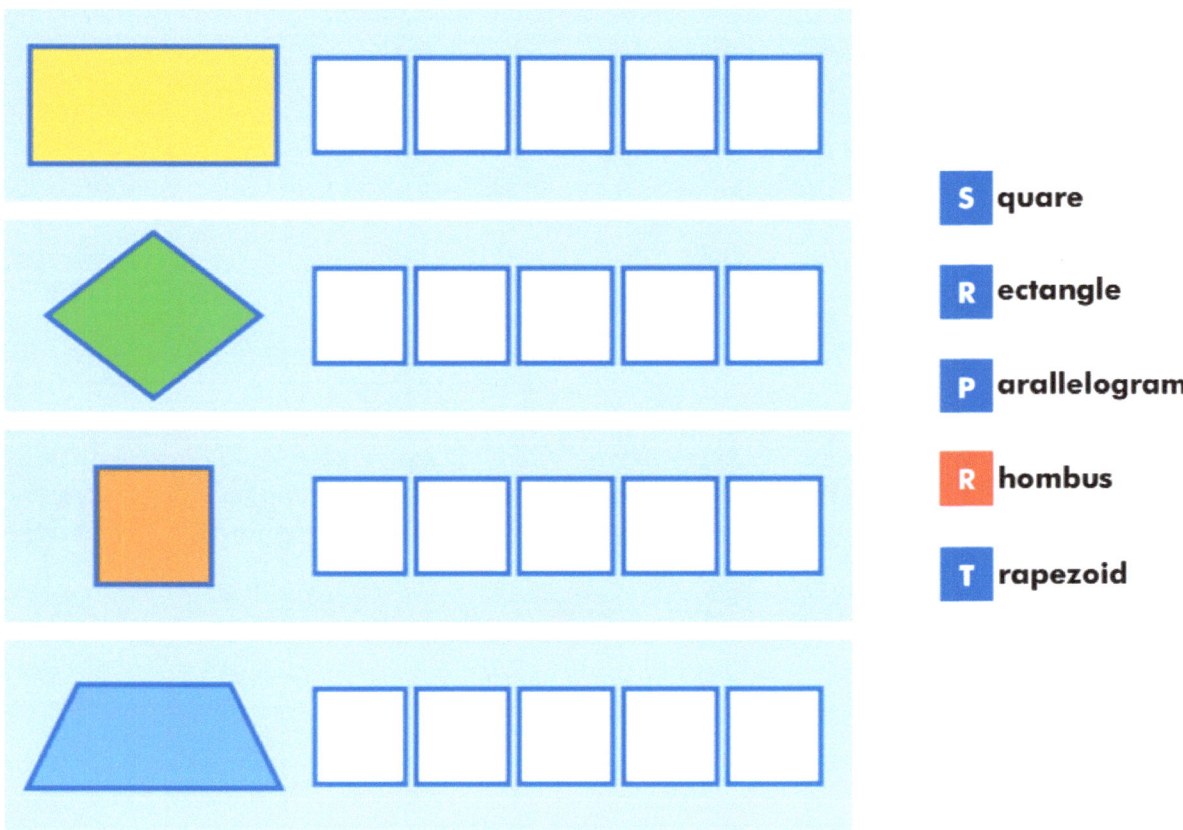

Name: _____

Properties of Quadrilaterals Quiz

1 True or false? A trapezoid is a parallelogram.

2 A square can be classified as which of the following?

   A Parallelogram

   B Rhombus

   C Rectangle

   D All of the above

3 True or false? A rectangle is always a parallelogram, but a parallelogram is not always a rectangle.

4 Three angles of a quadrilateral have the measures of 83°, 130°, and 95°. What is the measure of the 4th angle?

Triangle Classification

Key Vocabulary

equilateral triangle

isosceles triangle

acute angle

obtuse angle

right angle

What do you think the measurement of the third side of this triangle is? _____

An equilateral triangle

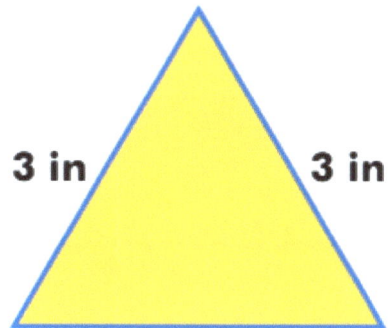

3 in   3 in

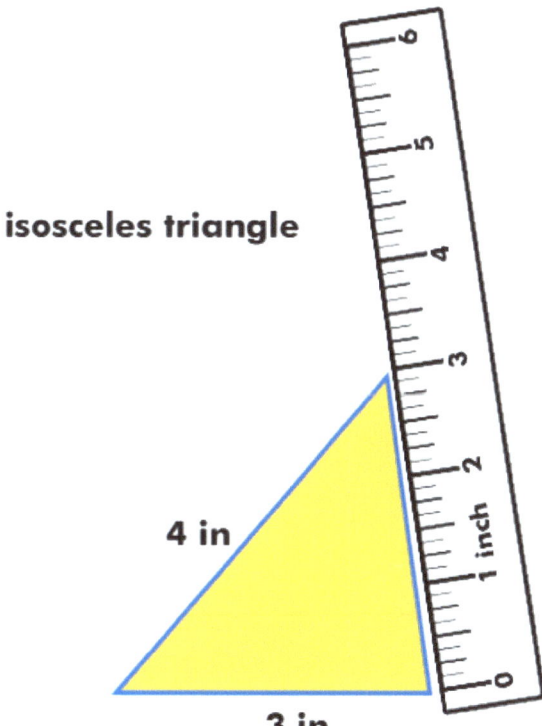

isosceles triangle

4 in

3 in

How is an isosceles triangle different from an equilateral triangle? _____

_____.

Draw a triangle with no congruent sides.
This is called a **scalene triangle**.

Connect the triangle facts with the correct triangle.

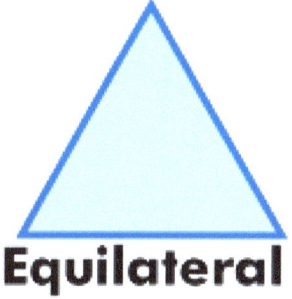

Equilateral

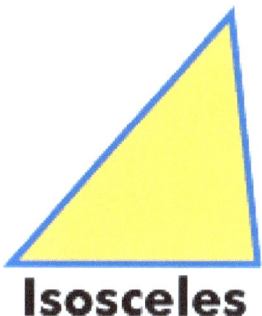

Isosceles

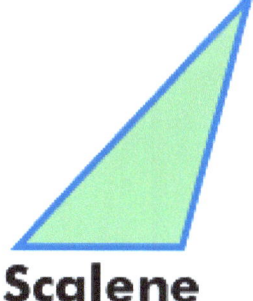

Scalene

Two angles of equal measure
No sides of equal length

Three sides of equal length
No angles of equal measure

Three angles of equal measure
Two sides of equal length

Classify these triangles by their sides.

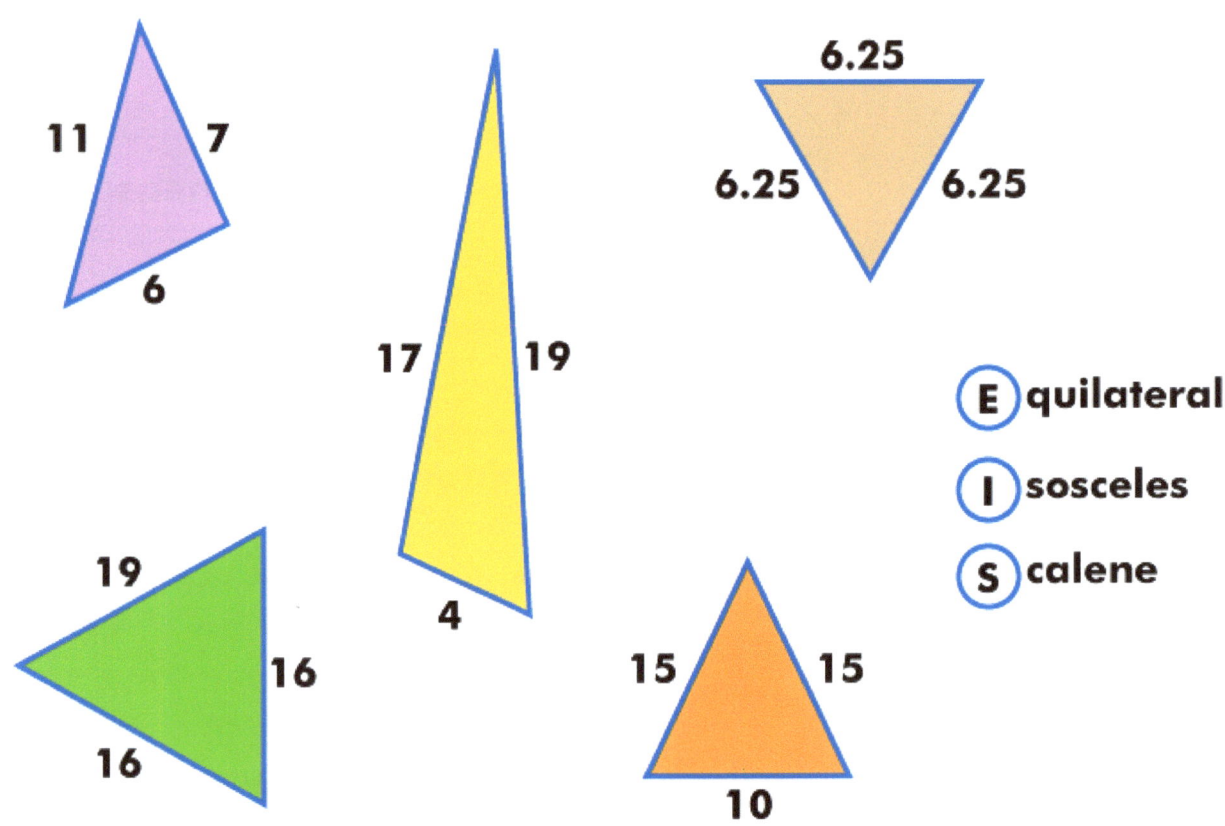

We can also classify triangles by the measure of their angles.

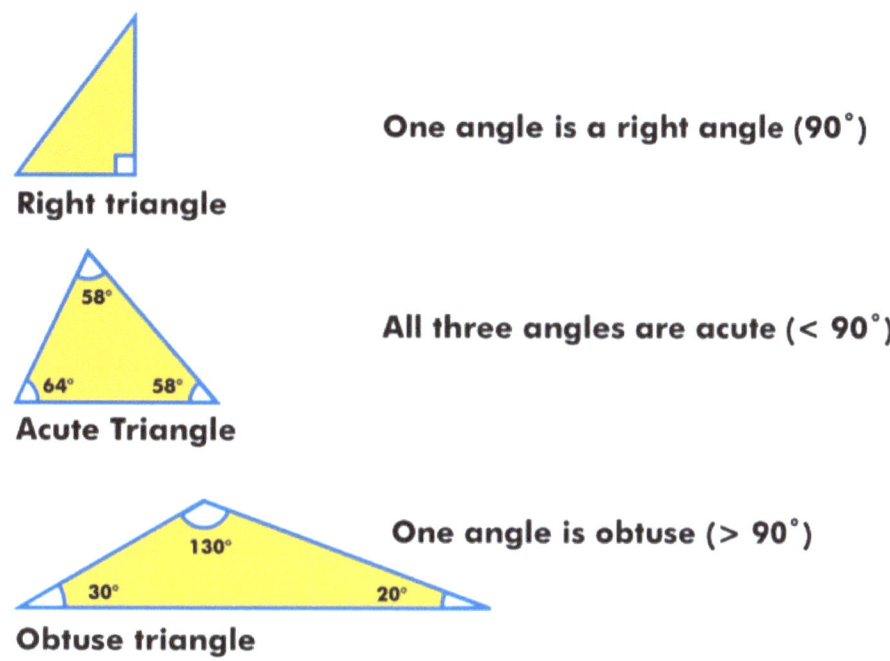

Right triangle — One angle is a right angle (90˚)

Acute Triangle — All three angles are acute (< 90˚)

Obtuse triangle — One angle is obtuse (> 90˚)

Draw these triangles.

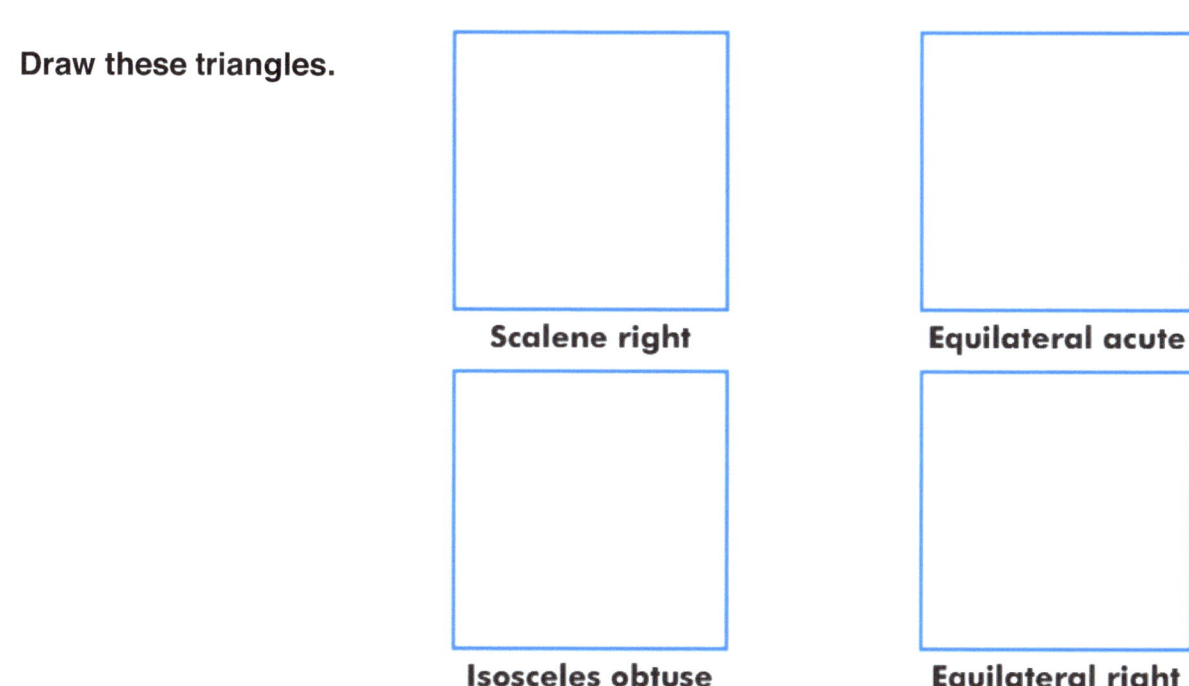

Scalene right

Equilateral acute

Isosceles obtuse

Equilateral right

Name: _____

Triangle Classification Quiz

1 True or false? Every equilateral triangle is an isosceles triangle.

2 True or false? Every isosceles triangle is an equilateral triangle.

3 Classify Figure 1.

- **A** Isosceles acute
- **B** Equilateral acute
- **C** Scalene acute
- **D** Scalene obtuse

4 Classify Figure 2.

- **A** Isosceles acute
- **B** Equilateral acute
- **C** Scalene acute
- **D** Scalene obtuse

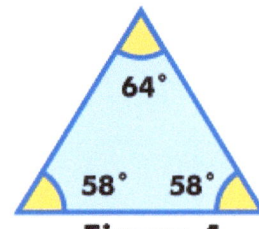

64°
58° 58°
Figure 1

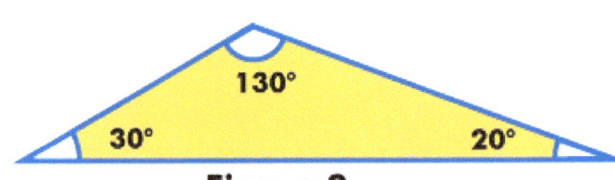

130°
30° 20°
Figure 2

Newburyport, MA 01950

1-800-596-3175

OnBoard Academics employs teachers to make lessons for teachers! We create and publish a wide range of aligned lessons in math, science and ELA for use on most EdTech devices including whiteboard, tablets, computers and pdfs for printing.

All of our lessons are aligned to the common core, the Next Generation Science Standards and all state standards.

If you like our products please visit our website for information on individual lessons, teachers licenses, building licenses, district licenses and subscriptions.

Thank you for using OnBoard Academic products.

www.ingramcontent.com/pod-product-compliance
Lightning Source LLC
Chambersburg PA
CBHW050408180526
45159CB00005B/2192